Graphic Organizers in Science™

Learning About the Water Cycle with Graphic Organizers

Isaac Nadeau

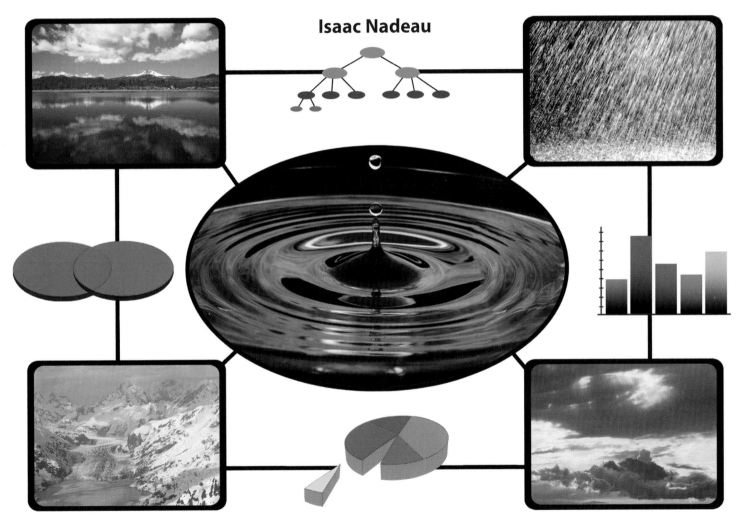

The Rosen Publishing Group's
PowerKids Press™
New York

Dedicated to Lake Monona

Published in 2005 by The Rosen Publishing Group, Inc.
29 East 21st Street, New York, NY 10010

First Edition

Editor: Natashya Wilson
Book Design: Mike Donnellan

Photo Credits: Cover (center) © Artville, LLC; cover (top left) © CORBIS; cover (top right) © Bruce Peebles/CORBIS; cover (bottom left) © Kennan Ward/CORBIS; cover (bottom right) © Image Club Graphics, Inc; p. 8 © Paul Seheult; Eye Ubiquitious/CORBIS; p 16 (right) © Galen Rowell/CORBIS; p. 19 © Yann Arthus-Bertrand/CORBIS.

Library of Congress Cataloging-in-Publication Data

Nadeau, Isaac.
Learning about the water cycle with graphic organizers / Isaac Nadeau.

    p. cm. — (Graphic organizers in science)
Summary: Uses texts and graphs to explain the water cycle on earth and its effects on life.
ISBN 1-4042-2808-X (lib. bdg.) – ISBN 1-4042-5046-8 (pbk.)
1. Hydrologic cycle—Study and teaching (Elementary)—Graphic methods—Juvenile literature. [1. Hydrologic cycle.] I. Title. II. Series.
GB848 .N324 2005
372.35'7—dc22

                            2003019158

Manufactured in the United States of America

Contents

Cycle: The Water Cycle

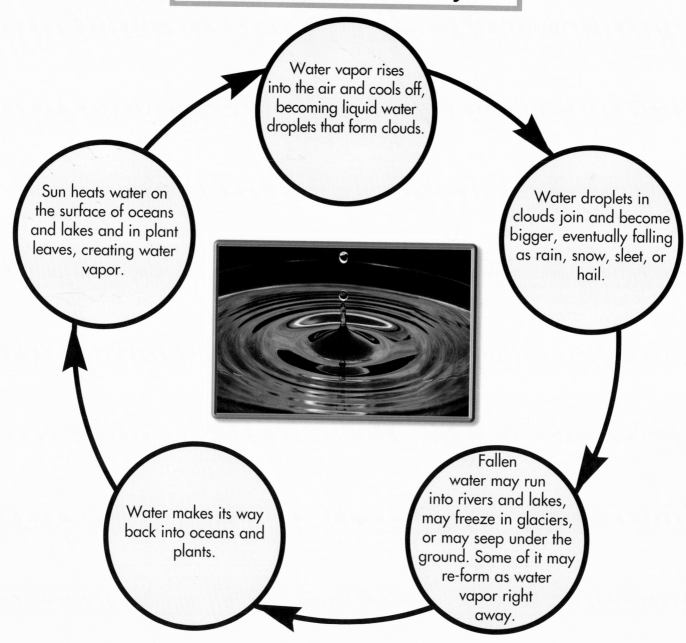

Water vapor rises into the air and cools off, becoming liquid water droplets that form clouds.

Sun heats water on the surface of oceans and lakes and in plant leaves, creating water vapor.

Water droplets in clouds join and become bigger, eventually falling as rain, snow, sleet, or hail.

Water makes its way back into oceans and plants.

Fallen water may run into rivers and lakes, may freeze in glaciers, or may seep under the ground. Some of it may re-form as water vapor right away.

What Is the Water Cycle?

All over Earth, water is moving. Whether water is falling as rain, moving as waves in the ocean, or filling up a bathtub, it is part of the water **cycle**. The water cycle is the process that moves water from one place to another on Earth. Water is always moving, from oceans and plants into the air, from the air into rivers, glaciers, or the ground, and then back into oceans and plants again. As water moves throughout the world, it changes from one form to another. Water can be a liquid, a solid, or a gas floating as tiny **droplets** of **water vapor** in the air. Each of the forms that water takes is an important part of the water cycle. All life on Earth depends on water and the water cycle.

In this book, **graphic organizers** are used to show facts about the water cycle. These visual tools help to make new ideas clear. You can use graphic organizers to study subjects in school.

This graphic organizer is called a cycle. Cycle organizers are used to show processes that happen over and over again. Each circle contains a step. The arrows lead from step to step. This cycle shows the basic steps of the water cycle.

Where in the World Is Water?

Water is found in many places. As a liquid, it can be either salt water or freshwater. About 97 percent of all water on Earth is salt water found in the oceans. Less than 3 percent of the water on Earth is freshwater. This is the water that people can drink. Freshwater is found in rivers, lakes, ponds, and other **wetlands**. It also collects underground. Freshwater is also found as snow and ice in the coldest places on Earth. There is more freshwater frozen in glaciers than there is in all other freshwater sources put together. **Brackish** water is found where freshwater flows into salt water. It has less salt than ocean water, but it is still salty. Water also moves through Earth's **atmosphere** as water vapor. Only a tiny amount of water is water vapor at any one time. The amount of water on Earth never changes. Water is always being **recycled** through the water cycle.

These graphic organizers are called pie charts. A pie chart compares amounts of the parts that make up a whole, or 100 percent. The top pie chart compares the water in Earth's oceans to the water not in oceans. The bottom chart shows where the water not in oceans is found.

Pie Chart: All the Water on Earth

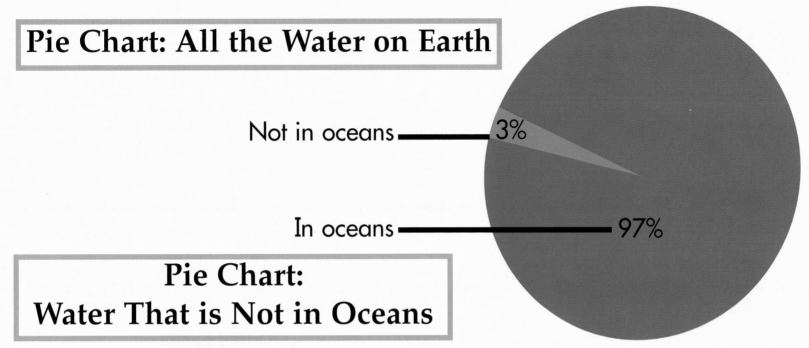

Not in oceans — 3%

In oceans — 97%

Pie Chart: Water That is Not in Oceans

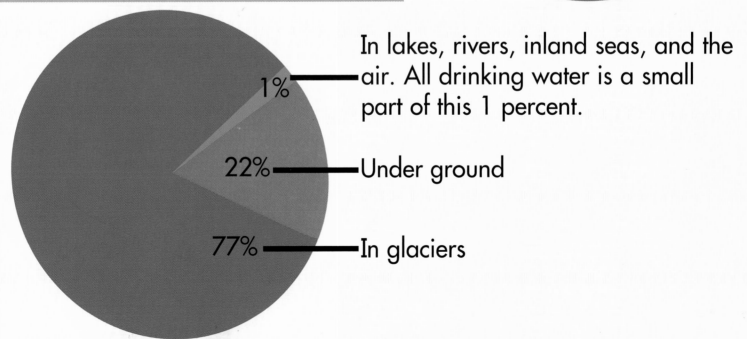

1% — In lakes, rivers, inland seas, and the air. All drinking water is a small part of this 1 percent.

22% — Under ground

77% — In glaciers

Concept Web: Evaporation

Wind

Wind, or moving currents of air, carries water vapor away from the surface of the water source, making room for more vapor to rise. The faster the wind blows, the faster the evaporation rate will be.

Temperature

As temperature rises, water evaporates faster. The more heat there is in the water, the more energy the water at the surface has and the faster it can evaporate.

Rate of Evaporation

Moisture

The amount of water vapor already in the air affects evaporation. If there is little moisture, more water vapor will rise. If there is already a lot of water vapor in the air, there is less room for more, so evaporation slows down.

Water on the Rise

The Sun is a major part of the water cycle. Heat from the Sun causes water on Earth's surface to change from a liquid into a gas called water vapor. This process is called **evaporation**. Heat also causes water to evaporate from plant leaves. This is called **transpiration**. Once water has evaporated, the water vapor floats in the air as droplets that are too small to be seen. These droplets are carried upward on currents of rising air.

The rate of evaporation depends on **temperature**, moisture that is already in the air, and wind. As temperature rises, more water evaporates. A rise of 18°F (10°C) makes water evaporate about twice as fast. However, if the air already has a lot of water vapor in it, evaporation slows down. Wind speeds up evaporation. It carries water vapor away with it, leaving dry air behind. More water vapor can rise into the dry air.

A concept web is a graphic organizer that shows the parts of a subject. In this concept web, the subject is the rate of evaporation. The three things that affect evaporation are moisture, temperature, and wind. They are placed around the subject.

Water Coming Down

All the water in the atmosphere will return to Earth's surface in the next step of the water cycle. As water vapor rises into the air, it becomes cooler. This cooling causes the water vapor droplets to gather and form larger droplets, returning to a liquid state. The process of water vapor cooling and becoming liquid is called **condensation**.

A cloud is a large group of water droplets gathered in cool air. These water droplets are still too light to fall to the ground. The air holds them up. As the air cools more, the water droplets continue to join. They form larger and larger droplets. Over time, these droplets become large and heavy enough to fall as rain or snow. When water falls to Earth, it is called **precipitation**.

Top: Water evaporates from lakes (left), condenses into clouds (middle), then falls as rain (right). These are steps in the water cycle. Bottom: This is a KWL chart. KWL charts are good ways to find out what you already know, what you want to know, and what you can learn by studying a subject. They can help you to decide what you need to study.

KWL Chart: Clouds and Rain

What I Know	What I Want to Know	What I Have Learned
• Rain is water that falls from the sky.	• How does water get into the sky?	• Heat from the Sun changes water on the surface of oceans and lakes and in plants. The water changes into water vapor and rises into the air. This is called evaporation.
• Clouds are in the sky.	• What are clouds made of and how do they form?	• As water vapor rises, the air around it gets colder. The vapor cools and begins to re-form as liquid water droplets. The droplets are so tiny and light that air holds them up. We see these droplets as clouds.
• Rain falls from clouds.	• Why does rain fall from clouds?	• The droplets in clouds keep joining together. Finally they become too big and heavy for the air to keep them up. The droplets fall, becoming rain.

Map: North America

U.S.A.

Atlantic Ocean

Mississippi River

Pacific Ocean

Gulf of Mexico

Mexico

Water On Land

When water falls to Earth's surface as precipitation, it can go many places as the water cycle continues. In hot places, such as deserts, water may evaporate right away and return into the atmosphere. In cooler places, water remains a liquid. Liquid water always flows downhill because of the pull of **gravity**. Rivers are examples of water flowing downhill. In cold places, water may stay on the ground as snow. It may melt when temperatures warm up. It may also become part of a glacier, where it may stay frozen for many years.

The movement of water over the ground is called **runoff**. A watershed is an area of land where runoff from many streams, lakes, and rivers flows toward a common place. Eventually gravity carries much of the water all the way to the ocean.

Maps are illustrations that show towns, countries, and even the whole world. They are used to find locations on Earth. This map shows rivers that flow into the Mississippi River. The smaller rivers carry runoff to the Mississippi, which carries the runoff to the gulf.

Water Down Below

When water falls onto land, some of it seeps into the ground. The process of water being **absorbed** by the ground is called **infiltration**. Water may seep only a few inches (cm) into the soil, where the roots of plants absorb it. Water may also seep many feet (m) below the surface. Water flowing underground has created caves by slowly carrying away bits of rock. Water also fills underground spaces called **aquifers**. An aquifer is a layer of soil or rock with air spaces in it that can be filled by water. When water fills the spaces, the aquifer is said to be **saturated**. The level of the water is called the **water table**. Water may stay underground for thousands of years, but someday it will reach the surface. It may become heated and rise to the surface as water vapor. It may flow underground until it reaches a river or ocean. Water may also flow up from a natural **spring**.

Top: *This diagram shows a lake formed at the level of the water table.* Bottom: *This graphic organizer is called a sequence chart. Sequence charts show the steps of a process in order. This chart shows the order in which water may move underground.*

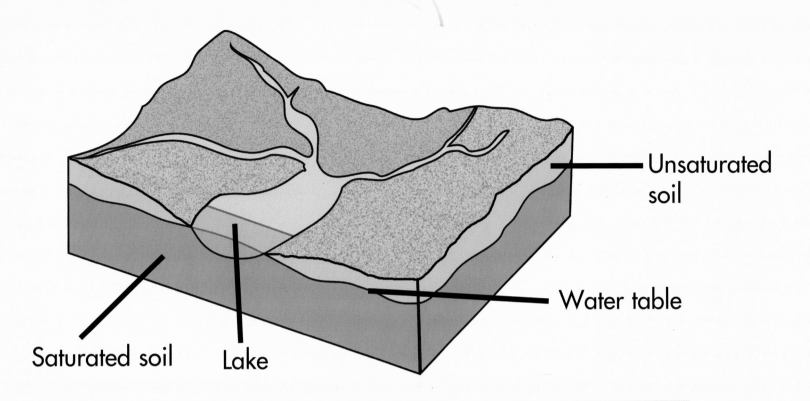

Unsaturated soil

Water table

Saturated soil

Lake

Sequence Chart: Water Under Ground

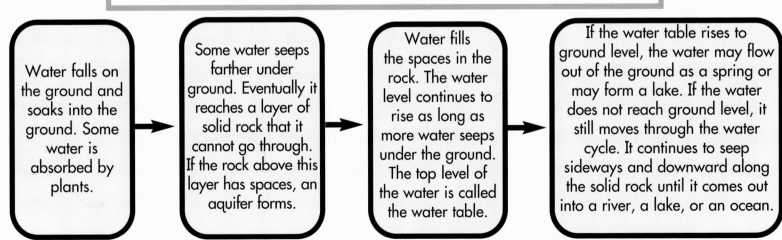

Water falls on the ground and soaks into the ground. Some water is absorbed by plants.

Some water seeps farther under ground. Eventually it reaches a layer of solid rock that it cannot go through. If the rock above this layer has spaces, an aquifer forms.

Water fills the spaces in the rock. The water level continues to rise as long as more water seeps under the ground. The top level of the water is called the water table.

If the water table rises to ground level, the water may flow out of the ground as a spring or may form a lake. If the water does not reach ground level, it still moves through the water cycle. It continues to seep sideways and downward along the solid rock until it comes out into a river, a lake, or an ocean.

Venn Diagram: Glaciers

Valley Glacier   Continental Glacier

- Forms within valleys high on mountains

- Movement is affected by the land's shape

- Can join with lower glaciers to form larger glaciers

- Melting ice at the end of the glacier may form a river

- Many exist today

- Shaped like a frozen river

- Formed by fallen snow that has piled up and packed into solid ice

- Temperature and precipitation affect movement and size

- Move at different rates, depending on the slope, thickness, and temperature of the ice

- Ice within glaciers also moves at different rates

- Covers the land

- Controls its own movement because it is so large

- Only two exist today, in Greenland and Antarctica

- Is thickest at the center and thinnest at the edges

Glaciers

Just as water may spend thousands of years under the ground as it moves through the water cycle, water can also spend thousands of years frozen as ice in glaciers. Glaciers are large bodies of ice that move downhill very slowly. They are found in places where a lot of snow falls in winter. Glaciers grow with the added snow of winter and shrink with the melting heat of summer. In summer glaciers do not melt completely.

There are two kinds of glaciers on Earth. They are valley glaciers and continental glaciers. Valley glaciers form in mountain valleys. Continental glaciers form in places where temperatures are cold year-round. The largest glacier on Earth covers the southern continent of Antarctica. This continental glacier, also called an ice sheet, is 3 miles (5 km) thick in some places.

Top: *A valley glacier* (left) *looks different from a continental glacier* (right). Bottom: *Venn diagrams are made of overlapping circles. They organize the features of related things so that you can learn how two things are alike and how they are different. The features that are alike go in the middle, where the circles overlap.*

Water and Climate

As it moves through the water cycle, water helps to create the world's climates. Climate is a usual pattern of weather, including precipitation and temperature, in a given place. Water creates rain, snow, and other moisture. Climates near the **equator**, the warmest area of Earth, are warm and rainy. Farther north and south, climates are cooler. Many places get snow in winter. The coldest climates are at the poles. The ice at the poles creates cold air that mixes with warm air from the equator, creating wind. Water also affects air temperature. Water takes longer than land to warm up and cool down. This keeps air temperatures above water more stable than air temperatures above land. When land is located near a large body of water, the air over the land mixes with the air over the water. The mix keeps the air temperature on land closer to the stable air temperature over the water.

A compare/contrast chart is another graphic organizer that you can use to compare things. This compare/contrast chart compares the climate of New York City, which is by the Atlantic Ocean, with that of Des Moines, Iowa, which is far inland.

Compare/Contrast Chart: New York and Des Moines		
	Average High and Low Temperatures in January	Average Precipitation in January
New York City, a Coastal Climate	High: 39°F (4°F) Low: 26°F (-3.3°C) Difference: 13°F (7.2°C)	3.3 inches (8.4 cm) of rain or snow
Des Moines, Iowa, an Inland Climate	High: 29°F (1.7°F) Low: 11°F (-11.7°C) Difference: 18°F (10°C)	1.1 inches (2.8 cm) of rain or snow

The cities of New York (left) and Des Moines are about the same distance north of the equator. However, Des Moines has a wider temperature difference and gets less precipitation. Des Moines is inland. New York is next to the Atlantic Ocean.

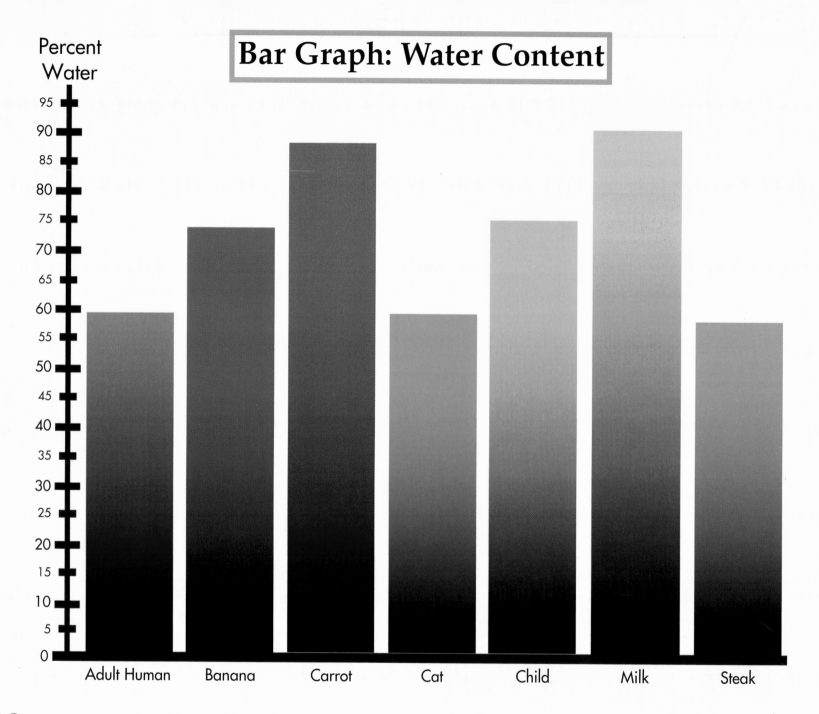

Water and Life

The water cycle is happening everywhere. Even the bodies of plants and animals take part in the water cycle. All living things on Earth need water. For instance, people need water to help break down food. The blood that carries oxygen and **nutrients** to all parts of the human body is mostly water. In fact, water makes up about 60 percent of the human body. Water passes from the human body through sweat and wastes. Plants also depend on water and are part of the water cycle. The water that plant roots soak up from the soil carries nutrients the plants need. The water keeps the plants firm and helps them to turn sunlight into energy. Water evaporates from plant leaves and rises into the sky, continuing the water cycle. Without the water cycle, water would not reach all of the living things on Earth that depend on it.

This is a bar graph. Bar graphs are used to compare things. Here, the percent of water in different subjects is compared. For example, an adult's body is about 60 percent water. The bar for the adult human goes up to 60 percent. Looking at the graph, you can see that a cat is also about 60 percent water. A child's body is about 75 percent water.

Water and People

Like all living things, people need water to live. People drink water and use water to grow food. People also use water for things such as bathing, sports, and moving goods by boat.

The amount of water on Earth has stayed the same for millions of years. Water is never used up. However, pollution can make water unusable for all living things. Many human activities cause pollution. Cars create a poisonous gas that rises in the air and ends up in rainwater. Some factories dump waste into water, making the water unclean. Sprays used to kill bugs on farm crops get washed into water when it rains, causing more pollution.

People are trying to find ways to keep water clean. Some people ride bikes or walk instead of driving cars when possible. Governments have passed laws to stop factories from dumping wastes into water. Learning about the water cycle can help people to keep water clean for all the living things on Earth.

Glossary

absorbed (ub-ZORBD) Taken in and held on to something.

aquifers (A-kwuh-furz) Layers of sand, gravel, or stone that hold water.

atmosphere (AT-muh-sfeer) The layers of air around Earth.

brackish (BRA-kish) Somewhat salty.

condensation (kon-den-SAY-shun) Cooled gas that has turned into drops of liquid.

cycle (SY-kul) A course of events that happens in the same order over and over.

droplets (DROP-lets) Tiny drops of liquid.

equator (ih-KWAY-tur) An imaginary line around Earth that separates it into two parts, northern and southern.

evaporation (ih-va-puh-RAY-shun) The process that changes a liquid, such as water, into a gas.

graphic organizers (GRA-fik OR-guh-ny-zerz) Charts, graphs, and pictures that sort facts and ideas and make them clear.

gravity (GRA-vih-tee) The natural force that causes objects to move toward the center of Earth.

infiltration (in-fil-TRAY-shun) Water soaking into the ground.

nutrients (NOO-tree-ints) Food that a living thing needs to live and to grow.

precipitation (preh-sih-pih-TAY-shun) Any moisture that falls from the sky.

recycled (ree-SY-kuld) Used again in a different way.

runoff (RUN-of) Water that flows into rivers after a rainstorm instead of going into the ground.

saturated (SA-chuh-rayt-ed) Completely filled with something, usually a liquid, such as water.

spring (SPRING) A natural fountain of water that flows out at Earth's surface.

temperature (TEM-pruh-cher) How hot or cold something is.

transpiration (tranz-puh-RAY-shun) Evaporation from the leaves and the stems of plants.

water table (WAH-ter TAY-bul) The level of water that collects in underground spaces in rock.

water vapor (WAH-ter VAY-pur) The gas state of water.

wetlands (WET-landz) Land with a lot of moisture in the soil.

Index

Web Sites

Due to the changing nature of Internet links, PowerKids Press has developed an online list of Web sites related to the subject of this book. This site is updated regularly. Please use this link to access the list:
www.powerkidslinks.com/gosci/watercyc/